RECRUITMENT, QUALIFICATION AND TRAINING OF PERSONNEL FOR NUCLEAR POWER PLANTS

The following States are Members of the International Atomic Energy Agency:

AFGHANISTAN
ALBANIA
ALGERIA
ANGOLA
ANTIGUA AND BARBUDA
ARGENTINA
ARMENIA
AUSTRALIA
AUSTRIA
AZERBAIJAN
BAHAMAS
BAHRAIN
BANGLADESH
BARBADOS
BELARUS
BELGIUM
BELIZE
BENIN
BOLIVIA, PLURINATIONAL STATE OF
BOSNIA AND HERZEGOVINA
BOTSWANA
BRAZIL
BRUNEI DARUSSALAM
BULGARIA
BURKINA FASO
BURUNDI
CAMBODIA
CAMEROON
CANADA
CENTRAL AFRICAN REPUBLIC
CHAD
CHILE
CHINA
COLOMBIA
COMOROS
CONGO
COSTA RICA
CÔTE D'IVOIRE
CROATIA
CUBA
CYPRUS
CZECH REPUBLIC
DEMOCRATIC REPUBLIC OF THE CONGO
DENMARK
DJIBOUTI
DOMINICA
DOMINICAN REPUBLIC
ECUADOR
EGYPT
EL SALVADOR
ERITREA
ESTONIA
ESWATINI
ETHIOPIA
FIJI
FINLAND
FRANCE
GABON
GEORGIA
GERMANY
GHANA
GREECE
GRENADA
GUATEMALA
GUYANA
HAITI
HOLY SEE
HONDURAS
HUNGARY
ICELAND
INDIA
INDONESIA
IRAN, ISLAMIC REPUBLIC OF
IRAQ
IRELAND
ISRAEL
ITALY
JAMAICA
JAPAN
JORDAN
KAZAKHSTAN
KENYA
KOREA, REPUBLIC OF
KUWAIT
KYRGYZSTAN
LAO PEOPLE'S DEMOCRATIC REPUBLIC
LATVIA
LEBANON
LESOTHO
LIBERIA
LIBYA
LIECHTENSTEIN
LITHUANIA
LUXEMBOURG
MADAGASCAR
MALAWI
MALAYSIA
MALI
MALTA
MARSHALL ISLANDS
MAURITANIA
MAURITIUS
MEXICO
MONACO
MONGOLIA
MONTENEGRO
MOROCCO
MOZAMBIQUE
MYANMAR
NAMIBIA
NEPAL
NETHERLANDS
NEW ZEALAND
NICARAGUA
NIGER
NIGERIA
NORTH MACEDONIA
NORWAY
OMAN
PAKISTAN
PALAU
PANAMA
PAPUA NEW GUINEA
PARAGUAY
PERU
PHILIPPINES
POLAND
PORTUGAL
QATAR
REPUBLIC OF MOLDOVA
ROMANIA
RUSSIAN FEDERATION
RWANDA
SAINT KITTS AND NEVIS
SAINT LUCIA
SAINT VINCENT AND THE GRENADINES
SAMOA
SAN MARINO
SAUDI ARABIA
SENEGAL
SERBIA
SEYCHELLES
SIERRA LEONE
SINGAPORE
SLOVAKIA
SLOVENIA
SOUTH AFRICA
SPAIN
SRI LANKA
SUDAN
SWEDEN
SWITZERLAND
SYRIAN ARAB REPUBLIC
TAJIKISTAN
THAILAND
TOGO
TONGA
TRINIDAD AND TOBAGO
TUNISIA
TÜRKİYE
TURKMENISTAN
UGANDA
UKRAINE
UNITED ARAB EMIRATES
UNITED KINGDOM OF GREAT BRITAIN AND NORTHERN IRELAND
UNITED REPUBLIC OF TANZANIA
UNITED STATES OF AMERICA
URUGUAY
UZBEKISTAN
VANUATU
VENEZUELA, BOLIVARIAN REPUBLIC OF
VIET NAM
YEMEN
ZAMBIA
ZIMBABWE

The Agency's Statute was approved on 23 October 1956 by the Conference on the Statute of the IAEA held at United Nations Headquarters, New York; it entered into force on 29 July 1957. The Headquarters of the Agency are situated in Vienna. Its principal objective is "to accelerate and enlarge the contribution of atomic energy to peace, health and prosperity throughout the world".

IAEA Safety Standards Series No. SSG-75

RECRUITMENT, QUALIFICATION AND TRAINING OF PERSONNEL FOR NUCLEAR POWER PLANTS

SPECIFIC SAFETY GUIDE

INTERNATIONAL ATOMIC ENERGY AGENCY
VIENNA, 2022

COPYRIGHT NOTICE

Marketing and Sales Unit, Publishing Section
International Atomic Energy Agency
Vienna International Centre
PO Box 100
1400 Vienna, Austria
fax: +43 1 26007 22529
tel.: +43 1 2600 22417
email: sales.publications@iaea.org
www.iaea.org/publications

Printed by the IAEA in Austria
October 2022
STI/PUB/2029

IAEA Library Cataloguing in Publication Data

Names: International Atomic Energy Agency.
Title: Recruitment, qualification and training of personnel for nuclear power plants / International Atomic Energy Agency.
Description: Vienna : International Atomic Energy Agency, 2022. | Series: IAEA safety standards series, ISSN 1020–525X ; no. SSG-75 | Includes bibliographical references.
Identifiers: IAEAL 22-01531 | ISBN 978–92–0–137422–6 (paperback : alk. paper) | ISBN 978–92–0–137222–2 (pdf) | ISBN 978–92–0–137322–9 (epub)
Subjects: Nuclear power plants. | Nuclear power plants — Employees — Recruiting. | Nuclear power plants — Employees — Training of. | Nuclear industry — Employees.
Classification: UDC 621.311.25:005.953 | STI/PUB/2029

FOREWORD

by Rafael Mariano Grossi
Director General

The IAEA's Statute authorizes it to "establish...standards of safety for protection of health and minimization of danger to life and property". These are standards that the IAEA must apply to its own operations, and that States can apply through their national regulations.

The IAEA started its safety standards programme in 1958 and there have been many developments since. As Director General, I am committed to ensuring that the IAEA maintains and improves upon this integrated, comprehensive and consistent set of up to date, user friendly and fit for purpose safety standards of high quality. Their proper application in the use of nuclear science and technology should offer a high level of protection for people and the environment across the world and provide the confidence necessary to allow for the ongoing use of nuclear technology for the benefit of all.

Safety is a national responsibility underpinned by a number of international conventions. The IAEA safety standards form a basis for these legal instruments and serve as a global reference to help parties meet their obligations. While safety standards are not legally binding on Member States, they are widely applied. They have become an indispensable reference point and a common denominator for the vast majority of Member States that have adopted these standards for use in national regulations to enhance safety in nuclear power generation, research reactors and fuel cycle facilities as well as in nuclear applications in medicine, industry, agriculture and research.

The IAEA safety standards are based on the practical experience of its Member States and produced through international consensus. The involvement of the members of the Safety Standards Committees, the Nuclear Security Guidance Committee and the Commission on Safety Standards is particularly important, and I am grateful to all those who contribute their knowledge and expertise to this endeavour.

The IAEA also uses these safety standards when it assists Member States through its review missions and advisory services. This helps Member States in the application of the standards and enables valuable experience and insight to be shared. Feedback from these missions and services, and lessons identified from events and experience in the use and application of the safety standards, are taken into account during their periodic revision.

I believe the IAEA safety standards and their application make an invaluable contribution to ensuring a high level of safety in the use of nuclear technology. I encourage all Member States to promote and apply these standards, and to work with the IAEA to uphold their quality now and in the future.

THE IAEA SAFETY STANDARDS

BACKGROUND

Radioactivity is a natural phenomenon and natural sources of radiation are features of the environment. Radiation and radioactive substances have many beneficial applications, ranging from power generation to uses in medicine, industry and agriculture. The radiation risks to workers and the public and to the environment that may arise from these applications have to be assessed and, if necessary, controlled.

Activities such as the medical uses of radiation, the operation of nuclear installations, the production, transport and use of radioactive material, and the management of radioactive waste must therefore be subject to standards of safety.

Regulating safety is a national responsibility. However, radiation risks may transcend national borders, and international cooperation serves to promote and enhance safety globally by exchanging experience and by improving capabilities to control hazards, to prevent accidents, to respond to emergencies and to mitigate any harmful consequences.

States have an obligation of diligence and duty of care, and are expected to fulfil their national and international undertakings and obligations.

International safety standards provide support for States in meeting their obligations under general principles of international law, such as those relating to environmental protection. International safety standards also promote and assure confidence in safety and facilitate international commerce and trade.

A global nuclear safety regime is in place and is being continuously improved. IAEA safety standards, which support the implementation of binding international instruments and national safety infrastructures, are a cornerstone of this global regime. The IAEA safety standards constitute a useful tool for contracting parties to assess their performance under these international conventions.

THE IAEA SAFETY STANDARDS

The status of the IAEA safety standards derives from the IAEA's Statute, which authorizes the IAEA to establish or adopt, in consultation and, where appropriate, in collaboration with the competent organs of the United Nations and with the specialized agencies concerned, standards of safety for protection of health and minimization of danger to life and property, and to provide for their application.

With a view to ensuring the protection of people and the environment from harmful effects of ionizing radiation, the IAEA safety standards establish fundamental safety principles, requirements and measures to control the radiation exposure of people and the release of radioactive material to the environment, to restrict the likelihood of events that might lead to a loss of control over a nuclear reactor core, nuclear chain reaction, radioactive source or any other source of radiation, and to mitigate the consequences of such events if they were to occur. The standards apply to facilities and activities that give rise to radiation risks, including nuclear installations, the use of radiation and radioactive sources, the transport of radioactive material and the management of radioactive waste.

Safety measures and security measures[1] have in common the aim of protecting human life and health and the environment. Safety measures and security measures must be designed and implemented in an integrated manner so that security measures do not compromise safety and safety measures do not compromise security.

The IAEA safety standards reflect an international consensus on what constitutes a high level of safety for protecting people and the environment from harmful effects of ionizing radiation. They are issued in the IAEA Safety Standards Series, which has three categories (see Fig. 1).

Safety Fundamentals

Safety Fundamentals present the fundamental safety objective and principles of protection and safety, and provide the basis for the safety requirements.

Safety Requirements

An integrated and consistent set of Safety Requirements establishes the requirements that must be met to ensure the protection of people and the environment, both now and in the future. The requirements are governed by the objective and principles of the Safety Fundamentals. If the requirements are not met, measures must be taken to reach or restore the required level of safety. The format and style of the requirements facilitate their use for the establishment, in a harmonized manner, of a national regulatory framework. Requirements, including numbered 'overarching' requirements, are expressed as 'shall' statements. Many requirements are not addressed to a specific party, the implication being that the appropriate parties are responsible for fulfilling them.

Safety Guides

Safety Guides provide recommendations and guidance on how to comply with the safety requirements, indicating an international consensus that it

---

[1] See also publications issued in the IAEA Nuclear Security Series.

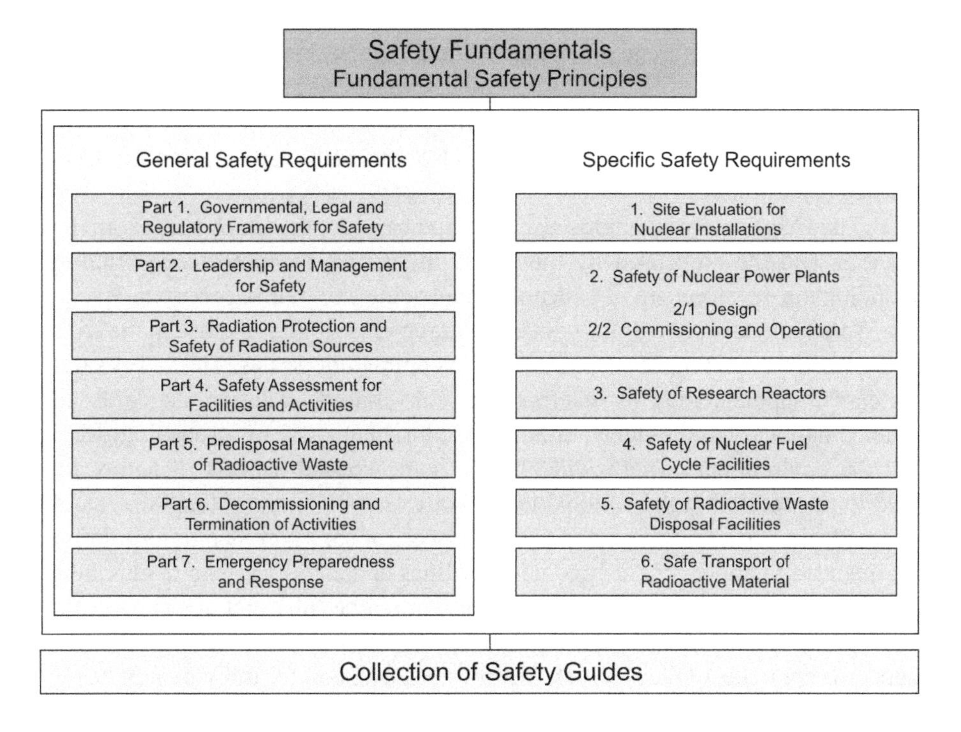

FIG. 1. The long term structure of the IAEA Safety Standards Series.

is necessary to take the measures recommended (or equivalent alternative measures). The Safety Guides present international good practices, and increasingly they reflect best practices, to help users striving to achieve high levels of safety. The recommendations provided in Safety Guides are expressed as 'should' statements.

APPLICATION OF THE IAEA SAFETY STANDARDS

The principal users of safety standards in IAEA Member States are regulatory bodies and other relevant national authorities. The IAEA safety standards are also used by co-sponsoring organizations and by many organizations that design, construct and operate nuclear facilities, as well as organizations involved in the use of radiation and radioactive sources.

The IAEA safety standards are applicable, as relevant, throughout the entire lifetime of all facilities and activities — existing and new — utilized for peaceful purposes and to protective actions to reduce existing radiation risks. They can be used by States as a reference for their national regulations in respect of facilities and activities.

The IAEA's Statute makes the safety standards binding on the IAEA in relation to its own operations and also on States in relation to IAEA assisted operations.

The IAEA safety standards also form the basis for the IAEA's safety review services, and they are used by the IAEA in support of competence building, including the development of educational curricula and training courses.

International conventions contain requirements similar to those in the IAEA safety standards and make them binding on contracting parties. The IAEA safety standards, supplemented by international conventions, industry standards and detailed national requirements, establish a consistent basis for protecting people and the environment. There will also be some special aspects of safety that need to be assessed at the national level. For example, many of the IAEA safety standards, in particular those addressing aspects of safety in planning or design, are intended to apply primarily to new facilities and activities. The requirements established in the IAEA safety standards might not be fully met at some existing facilities that were built to earlier standards. The way in which IAEA safety standards are to be applied to such facilities is a decision for individual States.

The scientific considerations underlying the IAEA safety standards provide an objective basis for decisions concerning safety; however, decision makers must also make informed judgements and must determine how best to balance the benefits of an action or an activity against the associated radiation risks and any other detrimental impacts to which it gives rise.

DEVELOPMENT PROCESS FOR THE IAEA SAFETY STANDARDS

The preparation and review of the safety standards involves the IAEA Secretariat and five Safety Standards Committees, for emergency preparedness and response (EPReSC) (as of 2016), nuclear safety (NUSSC), radiation safety (RASSC), the safety of radioactive waste (WASSC) and the safe transport of radioactive material (TRANSSC), and a Commission on Safety Standards (CSS) which oversees the IAEA safety standards programme (see Fig. 2).

All IAEA Member States may nominate experts for the Safety Standards Committees and may provide comments on draft standards. The membership of the Commission on Safety Standards is appointed by the Director General and includes senior governmental officials having responsibility for establishing national standards.

A management system has been established for the processes of planning, developing, reviewing, revising and establishing the IAEA safety standards.

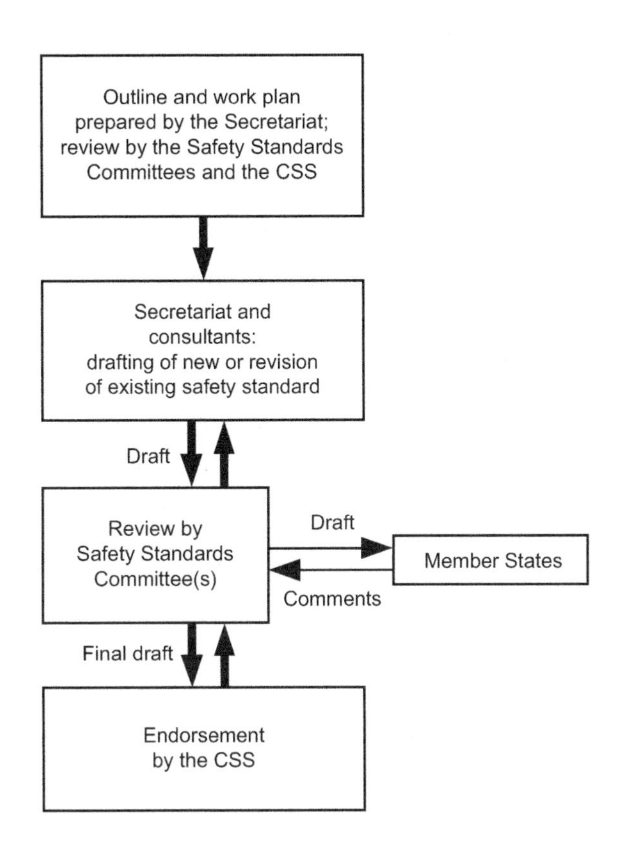

FIG. 2. The process for developing a new safety standard or revising an existing standard.

It articulates the mandate of the IAEA, the vision for the future application of the safety standards, policies and strategies, and corresponding functions and responsibilities.

INTERACTION WITH OTHER INTERNATIONAL ORGANIZATIONS

The findings of the United Nations Scientific Committee on the Effects of Atomic Radiation (UNSCEAR) and the recommendations of international expert bodies, notably the International Commission on Radiological Protection (ICRP), are taken into account in developing the IAEA safety standards. Some safety standards are developed in cooperation with other bodies in the United Nations system or other specialized agencies, including the Food and Agriculture Organization of the United Nations, the United Nations Environment Programme, the International Labour Organization, the OECD Nuclear Energy Agency, the Pan American Health Organization and the World Health Organization.

INTERPRETATION OF THE TEXT

Safety related terms are to be understood as defined in the IAEA Safety Glossary (see https://www.iaea.org/resources/safety-standards/safety-glossary). Otherwise, words are used with the spellings and meanings assigned to them in the latest edition of The Concise Oxford Dictionary. For Safety Guides, the English version of the text is the authoritative version.

The background and context of each standard in the IAEA Safety Standards Series and its objective, scope and structure are explained in Section 1, Introduction, of each publication.

Material for which there is no appropriate place in the body text (e.g. material that is subsidiary to or separate from the body text, is included in support of statements in the body text, or describes methods of calculation, procedures or limits and conditions) may be presented in appendices or annexes.

An appendix, if included, is considered to form an integral part of the safety standard. Material in an appendix has the same status as the body text, and the IAEA assumes authorship of it. Annexes and footnotes to the main text, if included, are used to provide practical examples or additional information or explanation. Annexes and footnotes are not integral parts of the main text. Annex material published by the IAEA is not necessarily issued under its authorship; material under other authorship may be presented in annexes to the safety standards. Extraneous material presented in annexes is excerpted and adapted as necessary to be generally useful.

CONTENTS

1. INTRODUCTION

BACKGROUND

1.1. Requirements for the operation of nuclear power plants are established in IAEA Safety Standards Series No. SSR-2/2 (Rev. 1), Safety of Nuclear Power Plants: Commissioning and Operation [1], while requirements for the design of nuclear power plants are established in IAEA Safety Standards Series No. SSR-2/1 (Rev. 1), Safety of Nuclear Power Plants: Design [2].

1.2. Requirements on establishing, sustaining and continuously improving leadership and management for safety and an effective management system for all facilities and activities are established in IAEA Safety Standards Series No. GSR Part 2, Leadership and Management for Safety [3].

1.3. This Safety Guide provides specific recommendations on the recruitment, qualification and training of personnel for nuclear power plants to ensure the safe operation of the nuclear power plant.

1.4. This Safety Guide was developed in parallel with six other Safety Guides on the operation of nuclear power plants, as follows:

- IAEA Safety Standards Series No. SSG-70, Operational Limits and Conditions and Operating Procedures for Nuclear Power Plants [4];
- IAEA Safety Standards Series No. SSG-71, Modifications to Nuclear Power Plants [5];
- IAEA Safety Standards Series No. SSG-72, The Operating Organization for Nuclear Power Plants [6];
- IAEA Safety Standards Series No. SSG-73, Core Management and Fuel Handling for Nuclear Power Plants [7];
- IAEA Safety Standards Series No. SSG-74, Maintenance, Testing, Surveillance and Inspection in Nuclear Power Plants [8];
- IAEA Safety Standards Series No. SSG-76, Conduct of Operations at Nuclear Power Plants [9].

A collective aim of this set of Safety Guides is to support the fostering of a strong safety culture in nuclear power plants.

1.5. The terms used in this Safety Guide are to be understood as defined and explained in the IAEA Safety Glossary [10].

1.6. This Safety Guide supersedes IAEA Safety Standards Series No. NS-G-2.8, Recruitment, Qualification and Training of Personnel for Nuclear Power Plants[1].

OBJECTIVE

1.7. The purpose of this Safety Guide is to provide recommendations on the recruitment, qualification and training of personnel for nuclear power plants to meet the requirements established in SSR-2/2 (Rev. 1) [1], in particular Requirements 4, 7 and 18.

1.8. The recommendations provided in this Safety Guide are aimed primarily at operating organizations of nuclear power plants and regulatory bodies.

SCOPE

1.9. It is expected that this Safety Guide will be used primarily for land based stationary nuclear power plants with water cooled reactors designed for electricity generation or for other production applications (such as district heating or desalination).

1.10. This Safety Guide identifies the main objectives and responsibilities of the operating organization for the recruitment, qualification and training of personnel for new and existing nuclear power plants to establish and maintain a high level of competence of personnel and to ensure safe operation of the nuclear power plant.

1.11. This publication can also be used as a guide for the recruitment, training and qualification of personnel for nuclear installations other than nuclear power plants and also for nuclear power plants other than water cooled nuclear power plants.

STRUCTURE

1.12. Recommendations relating to the recruitment and selection of suitable personnel for a nuclear power plant are provided in Section 2. Section 3 provides recommendations on the establishment of specifications for competence and

---

[1] INTERNATIONAL ATOMIC ENERGY AGENCY, Recruitment, Qualification and Training of Personnel for Nuclear Power Plants, IAEA Safety Standards Series No. NS-G-2.8, IAEA, Vienna (2002).

qualifications, including the educational background and experience of the personnel, for tasks relating to safety. Section 4 provides recommendations on the development of the training policy to ensure safe operation of the plant and addresses the systematic approach to training, training settings and methods, initial and continuing training, documentation of training and training programmes for emergencies. Section 5 provides recommendations on the main aspects of training programmes, including specific considerations for different categories of personnel. Section 6 provides recommendations on training facilities and materials. Section 7 provides recommendations on authorizing personnel to perform certain tasks or duties with a direct bearing on safety.

2. RECRUITMENT AND SELECTION OF PERSONNEL FOR NUCLEAR POWER PLANTS

STAFFING ARRANGEMENTS FOR NUCLEAR POWER PLANTS

2.1. Requirement 4 of SSR-2/2 (Rev. 1) [1] states that "**The operating organization shall be staffed with competent managers and sufficient qualified personnel for the safe operation of the plant.**"

2.2. The staffing of the plant may be supplemented, as necessary, by consultants or contractors to ensure that duties relevant to safety can be performed without undue haste or pressure.

2.3. Paragraph 3.11 of SSR-2/2 (Rev. 1) [1] states that "A long term staffing plan aligned to the long term objectives of the operating organization shall be developed in anticipation of the future needs of the operating organization for personnel and skills."

2.4. The staffing plan should be regularly reassessed and updated to reflect organizational changes. Such changes can result from changes to work programmes, the adoption of new technology, the addition of reactor units or from changes in stages of the lifetime of the plant. Organizational changes might also occur as a result of feedback of operating experience (see Requirement 24 of SSR-2/2 (Rev. 1) [1]), especially when significant improvements in safety or in the understanding of root causes can help to avoid the recurrence of events. Issues such as the age profiles of plant personnel, advances in automatic control and changes in waste management policies should also initiate a reassessment of the staffing plan. Fluctuations in staffing and staff motivation (e.g. for plants facing shutdown) should be taken into account when performing the periodic reassessments of the staffing plan.

2.5. Paragraph 3.13 of SSR-2/2 (Rev. 1) [1] states:

> "A staff health policy shall be instituted and maintained by the operating organization to ensure the fitness for duty of personnel. Attention shall be paid to minimizing conditions causing stress, and to setting restrictions on overtime and setting requirements for rest breaks. The health policy shall cover the prohibition of alcohol consumption and drug abuse."

The prohibition of alcohol consumption and drug abuse should be strictly enforced. A programme to identify alcohol consumption and drug abuse should be established.

2.6. Further recommendations on organizational arrangements for the staffing of nuclear power plants are provided in SSG-72 [6].

RECRUITMENT POLICY

2.7. Paragraph 3.11 of SSR-2/2 (Rev. 1) [1] states that "The recruitment and selection policy of the operating organization shall be directed at retaining competent personnel to cover all aspects of safe operation." The operating organization is responsible for the recruitment and selection of staff. Suitably qualified personnel should be selected and recruited in accordance with approved procedures. Attitude towards safety should be a selection criterion in the recruitment of plant personnel (see para. 2.16).

2.8. Paragraph 3.10 of SSR-2/2 (Rev. 1) [1] states:

> "The operating organization shall be responsible for ensuring that the necessary knowledge, skills, attitudes and safety expertise are sustained at the plant, and that long term objectives for human resources policy are developed and are met."

The recruitment and selection policy at a nuclear power plant should be aimed at retaining a pool of experienced staff with a broad range of operational and safety expertise.

2.9. Staff motivation and career development should be considered in the recruitment and selection processes. Promoting personnel within the operating organization ensures that high quality work is rewarded and provides a motivation for personnel to enhance their competence to enable them to apply for higher positions. Job stability is also an asset that the operating organization should use to sustain staff motivation while finding the right balance between internal promotion and external recruitment.

2.10. It might be necessary to hire personnel externally if positions at the plant cannot be filled internally. Candidates with relevant qualifications and experience may be recruited from conventional power plants, design groups and nuclear research establishments, and may be given appropriate practical experience and training at a nuclear power plant under the guidance of experienced staff. Candidates from conventional power plants should be given due consideration for recruitment to direct operations and maintenance positions, because of their experience in these areas and their ability to cope with the needs of day-to-day operations.

2.11. When it is not possible to recruit individuals with the necessary experience, consideration should be given to recruit personnel directly from schools, technical colleges and universities. Specialized training should then be planned and provided, including on the job training on specific systems and equipment and simulator training at the plant and at other organizations, as appropriate.

2.12. The operating organization should plan for the recruitment and selection of personnel for a new plant (i.e. in terms of positions and recruitment schedules) before the start of plant construction. The replacement of personnel who reach retirement age at an operating plant and the appointment of personnel to decommission a plant should also be anticipated, in order to prepare recruitment and selection schedules. However, vacancies might also arise through personnel leaving or moving to other positions, or through premature retirement. Such situations necessitate some flexibility in the recruitment and selection processes described in this section.

2.13. Medical fitness for duty expectations should be clearly defined for each position. The operating organization should ensure that all operating personnel whose duties have a bearing on safety are medically examined at the time of recruitment (and periodically thereafter) to ensure that their state of health is suitable for the duties and responsibilities assigned to them. Aptitude tests should be used, where appropriate. For certain positions, the operating organization may also conduct psychological tests. All personnel who might be occupationally exposed to radiation at the plant should be subject to initial and periodic medical examinations, as appropriate.

SELECTION OF CANDIDATES

2.14. The selection of candidates for vacant positions should be based on a candidate's potential to develop the necessary competence, through additional training, experience and development. A candidate's potential for occupying higher positions may also be taken into account.

2.15. The selection process should include the following steps:

(a) Establishing criteria (including medical criteria — see para. 2.13) for accepting or rejecting applications and for classifying acceptable candidates;
(b) Obtaining information about candidates, including security information, in accordance with any relevant regulatory requirements;
(c) Interviewing candidates;
(d) Objective testing of candidates to measure their aptitude for the job;
(e) Assessing candidates against the criteria in point (a) in order to reach a hiring decision.

The selection criteria should be based on the job specification, and the necessary entry level qualifications and competences developed through education, experience and prior training should be clearly stated (see Section 3).

2.16. The selection criteria should include such factors as qualifications, experience, problem solving ability, emotional stability, motivation, initiative, communication skills, work background and, where appropriate, physical aptitude for specific tasks. In addition, specific attributes relating to safety culture, such as a questioning attitude, a rigorous and prudent approach, and communication and learning abilities, should be taken into consideration in selecting candidates at a nuclear power plant.

2.17. When selecting candidates to work as control room operators or as other personnel who might have to respond to an emergency, their ability to work together as a team in such conditions should be considered. In the allocation of staff to particular teams, likely personal interactions should be taken into account.

2.18. Some recruits might need extended periods of time for gaining additional experience, training or even further education to reach the level of competence necessary for the position they will eventually occupy. The selection process should identify those recruits who are prepared to learn and be trained, who meet the entry criteria for the relevant courses, and who are likely to achieve the necessary competence.

2.19. Candidates for managerial and supervisory positions should be selected on the basis of criteria that include appropriate attitudes to safety and that emphasize proven, conservative, safety enhancing decision making skills. The following factors should be taken into account:

(a) Management skills, including analytical, supervisory, leadership and communication abilities;
(b) Experience (performance in previous jobs);
(c) Education and training;
(d) Knowledge of plant operations;
(e) Psychological and medical criteria;
(f) Attitudes towards nuclear safety;
(g) Attitudes towards learning and self-learning;
(h) Attitudes towards the training and career development of personnel.

2.20. The final selection of a candidate should be based on the recommendations of a group of skilled and knowledgeable staff who are familiar with the competence needed and with the human resources policies of the operating organization. Where no candidate is found who meets the criteria, a structured process should be applied to determine appropriate temporary arrangements for covering vacant positions.

RECRUITMENT PROGRAMME FOR NEW NUCLEAR POWER PROGRAMMES

2.21. An operating organization planning to build a first nuclear power plant (or the first of a new type of plant with significant technological differences to existing plants) should begin its recruitment programmes earlier than those operating organizations that already have experience with such a plant and can call upon an existing pool of expertise. Initial recruitment should be completed in sufficient time before the commissioning of a plant to allow personnel to gain experience of the design, construction and commissioning stages by working alongside contractors and commissioning personnel, and to receive appropriate training and familiarize themselves with the plant. By participating in these stages, personnel will acquire a better understanding of the design intents, the assumptions on which the safety criteria are based and the technical characteristics of the plant. Further information on recruitment programmes can be found in IAEA Nuclear Energy Series No. NG-G-3.1 (Rev. 1), Milestones in the Development of a National Infrastructure for Nuclear Power [11] and in IAEA Safety Standards Series No. SSG-16, Establishing the Safety Infrastructure for a Nuclear Power Programme [12].

2.22. The operating organization should prepare a schedule showing how the initial recruitment and selection of personnel will be planned and implemented. If a plant is the first of several of a type to be built, the period covered by this schedule should begin with the start of the construction work. As additional plants are constructed and operated, this period might be reduced.

2.23. Senior managers, particularly the plant manager and deputy plant manager, should be recruited at the very beginning of the recruitment period, since their first duties include supporting preoperational activities and planning and conducting further recruitment and training. Particular attention should be given to the early recruitment of shift supervisors and senior reactor operators.

2.24. Operators, engineers and technicians for a new plant should be recruited early. Time should be allocated for performing duties for which no additional training is necessary, or that can be accomplished under supervision before commissioning. Time should also be allocated to operators, engineers and technicians for training before they assume the full responsibilities of their positions. Some of these personnel could be assigned to the plant supplier or construction organization to obtain experience with new and/or complex equipment.

2.25. The training of technicians should include working with the construction organization and commissioning personnel in the checking and initial calibration of instruments and controls, and in the initial operation of such equipment before fuel loading. This also provides a valuable opportunity for personnel to familiarize themselves with parts of the plant that will subsequently have limited or no access. Some individuals might therefore be recruited and employed at the site well before the commencement of operation.

2.26. The initial recruitment of experienced personnel with specific manual skills needed for a new plant should start in such a way that these people can discharge their duties according to the specified safety and quality requirements. Such persons might receive specialized training from suppliers during the assembly and pre-shipment testing of special equipment such as diesel or gas turbine sets, large pumps or fuel handling equipment. Some of them might need lengthy training programmes if particular skills asked of them are as yet unavailable to the operating organization, such as those of a certified nuclear welder. Recruitment should continue in accordance with operational needs.

2.27. Further information on workforce planning for new nuclear power programmes is provided in Ref. [13].

3. COMPETENCE AND QUALIFICATION OF NUCLEAR POWER PLANT PERSONNEL

3.1. Requirement 7 of SSR-2/2 (Rev. 1) [1] states that "**The operating organization shall ensure that all activities that may affect safety are performed by suitably qualified and competent persons.**" Furthermore, para. 4.16 of SSR-2/2 (Rev. 1) [1] states:

> "The operating organization shall clearly define the requirements for qualification and competence to ensure that personnel performing safety related functions are capable of safely performing their duties. Certain operating positions may require formal authorization or a licence."

Competence is the ability to apply skills, knowledge and attitudes in order to perform an activity or a job to a specified level in an effective and efficient manner. Competence may be developed through education, experience and formal vocational training. Qualification is a formal statement resulting from an assessment or audit of an individual's competence to fill a position and perform all duties assigned to that position in a responsible manner. The criteria for competence and qualification should be established in such a way as to ensure that the competences are appropriate to the tasks and activities to be performed. The operating organization should determine positions for which a formal authorization is necessary before an individual is assigned to a designated position (see Section 7).

3.2. The operating organization should ensure that all personnel who perform duties that affect safety have a sufficient understanding of the plant and its safety features and sufficient other competences (e.g. leadership, management and supervisory skills, and 'soft skills' such as teamworking and communication) to perform their duties safely. All such personnel should be trained in safety management in their areas of responsibility and in accordance with their assigned duties and tasks.

3.3. Before undertaking any safety related work, personnel should demonstrate the appropriate knowledge, skills and attitudes to ensure safety under a variety of conditions relating to their duties. Such personnel should be trained in how to promote safety culture, including a rigorous and prudent approach to safety, a questioning attitude and conservative decision making (see para. 4.30 of SSR-2/2 (Rev. 1) [1]).

3.4. When personnel are to be replaced, a reasonable overlap should be arranged to enable knowledge transfer between the outgoing and incoming personnel so that the new personnel can acquire an understanding of their duties and responsibilities and of ongoing activities before assuming their positions.

3.5. The competence of each individual should be assessed against established criteria before that individual is assigned to a position. The competence of all individuals should be reassessed periodically while they perform their duties in the workplace.

3.6. Appropriate records of assessments against criteria for competence and qualification should be established and maintained for each individual at the plant.

3.7. The functions and the related duties and responsibilities of qualified personnel should be clearly indicated in the structure of the operating organization and in the job description for each position. For each category of personnel, the necessary competence may be defined by means of the following:

(a) Educational level (academic qualification);
(b) Previous experience (including direct and related experience);
(c) Initial training and continuing training.

3.8. Irrespective of any formal qualifications or certificates issued by other bodies, it is the responsibility of the operating organization to ensure the appropriate qualification of all relevant personnel. The responsibility of ensuring that individuals remain appropriately qualified also rests with the operating organization, although individuals should accept some responsibility in maintaining and developing their own competence through continuing professional training.

3.9. The need for specific skills and knowledge will be different for different positions. The balance between managerial competences and technical competences should be evaluated by the operating organization in establishing qualification criteria.

EDUCATIONAL BACKGROUND

3.10. Education provides the general knowledge and develops the attitudes, behaviours and intellectual skills that are the foundations of competence. Appropriate criteria for educational background should be established for all positions at the plant. These criteria should be taken into consideration in the preparation of training programmes for plant personnel. In turn, training programmes should be used to complement formal education with practical and job related knowledge and skills.

3.11. The operating organization should conduct an analysis of the knowledge and skills developed through the national education system, to help decide which educational qualifications are necessary for each particular position at its plant.

3.12. Operating personnel should have a sufficient basic knowledge of the main topics that affect plant operation, including radiation protection and nuclear safety. The plant manager, heads of departments, shift supervisors and reactor operators should have an educational background in engineering, reactor physics or nuclear technology.

3.13. The scope of knowledge, and therefore the criteria for educational background, should be commensurate with the position to be occupied. Managers and technical specialists should possess a wide knowledge of general science and technology (physics, mathematics, chemistry, thermodynamics). Managers, shift supervisors and operators should have knowledge of nuclear science and nuclear engineering. Managers and technical specialists should also have in depth knowledge of the specific areas relating to their work (e.g. of mechanical, electrical, electronic, chemical or civil engineering). Individuals recruited for managerial positions should additionally have an educational background in administration and human resources management.

3.14. Managers and technical specialists should have a university degree or equivalent certification in management, engineering or science, or some other educational background appropriate to the national education system and the specific job assigned. They might also have attained the necessary competence through appropriate experience and training, where this is permitted by the national regulations. Many positions, however, should be filled by individuals with formal educational qualifications. For example, the title of electrical engineer is established by the awarding of a formal degree and cannot be obtained through experience or training.

3.15. As a general rule, the education of technicians should provide them with the following:

(a) A good understanding of the basic sciences that form the foundation of the area of technology in which they will be working;
(b) Detailed knowledge of their area of technology;
(c) Practical training and experience in the application of their knowledge and skills;
(d) The capability to communicate, which includes oral, written and technical communication.

3.16. For positions that involve specific manual skills and corresponding knowledge, individuals can acquire these in vocational schools or apprenticeship programmes.

3.17. The following practices in relation to the educational background of nuclear power plant personnel are commonly applied [14]:

(a) Managerial positions (e.g. plant manager, deputy plant manager, operations manager, safety manager, maintenance manager, quality assurance manager, technical support manager, training manager) are usually occupied by university graduates in engineering or physical sciences.
(b) The other positions for which a university degree is normally expected are those of shift technical adviser and safety engineer. Reactor physicists, radiation protection officers, plant chemists and maintenance engineers will also generally have university degrees, and some of the more junior personnel might also have completed university level education.
(c) Supervisors (e.g. for the plant, unit, shift and control room) will often have a degree from a university or engineering college. Control room operators are typically expected to have a diploma from a technical school, although they might instead have a degree from a university or engineering college.
(d) Field operators[2] commonly have, at a minimum, a secondary school diploma.
(e) Other technical positions might be filled by graduates of vocational or technical schools.

---

[2] Field operators are the operating personnel assigned to control operational activities outside the control room.

3.18. Training instructors should have an appropriate background in an education related subject, in addition to a degree in an appropriate discipline in their area(s) of responsibility (see also para. 4.23 of SSR-2/2 (Rev. 1) [1]).

WORK EXPERIENCE

3.19. Experience is the knowledge gained and the skills developed while performing the duties of a position. Three principal grades of experience can be distinguished as follows:

(a) General plant experience, which comprises a general knowledge of nuclear power plants and their related activities. This sort of experience may be gained by occupying various positions at different plants.
(b) Plant familiarity, which is the detailed knowledge of a particular plant or activity, and which can only be obtained through day-to-day work in a particular position.
(c) Breadth of experience, which relates to knowledge not directly connected with the duties of a particular position. It includes a knowledge of interfacing activities, and a wider knowledge of the plant and the operating organization, which might extend to other activities outside the plant.

3.20. General plant experience provides broadly applicable knowledge of the properties of the plant (or of maintenance or similar activities). This type of experience can be acquired by working at different plants. Working in several plants can add to the general plant experience of operating personnel, and performing maintenance activities in different types of plant adds to the general plant experience of maintenance personnel. By performing these tasks and duties in operating plants, a knowledge of plant behaviour is accumulated over a period of time, which can be applied generally to a range of plants. Operators who need formal authorization (see para. 3.1) should have sufficient general plant experience, of which a part might have been acquired in other plants, including in conventional power plants. This type of knowledge appears to be retained for long periods, even after work at a plant has ended.

3.21. Plant familiarity can only be acquired and maintained by working at a plant in a certain position. Familiarity with the plant is essential for control room operators and other operators, to ensure that they can recall details of the plant within the time limits allowed for decisions and actions. This type of knowledge is lost rapidly after regular work in the position has ceased. Paragraph 4.19 of SSR-2/2 (Rev. 1) [1] states:

> "The training programme shall include provision for periodic confirmation of the competence of personnel and for refresher training on a regular basis. The refresher training shall also include retraining provision for personnel who have had extended absences from their authorized duties."

3.22. Breadth of experience comprises knowledge of matters beyond the activities relating to a specific position. Many operating organizations have adopted job rotation on a regular basis as a means of ensuring that operating personnel develop broad experience.

3.23. Experience in safety management and the development of safety culture at a nuclear power plant (or other relevant nuclear facility) is important for managers and personnel. The minimum amount of such experience needed for managers and operating personnel at a nuclear power plant should be specified by the operating organization.

3.24. For all positions at a nuclear power plant, some experience may be gained in design, construction and commissioning activities. The aim should be to develop an understanding of the design intents and assumptions, of the safety criteria and of the technical characteristics of the plant.

3.25. Examples of the length and breadth of experience necessary for specific types of position at a nuclear power plant are given in paras 3.26–3.34; the experience needed in individual States might differ. The number of years should not be given as much importance as the quality of the experience, the competence of the organization from which that experience was gained, and the level of individual responsibility. Documented recommendations of educators, instructors and former employers should be assigned considerable importance in evaluating the experience and competence of an individual. Individuals who have previously held an authorization for a significant length of time from an operating organization for a nuclear power plant should be considered to have documented experience. For plant managers, supervisors and control room operators, documented experience might be the best indicator of future work performance and safe operation.

Work experience for managers and supervisors at a nuclear power plant

3.26. Successful performance in subordinate positions is an acceptable form of experience for assignment to a senior position at a nuclear power plant. The plant manager, deputy plant manager, operations manager, safety manager, maintenance manager, quality assurance manager, technical support manager and training manager should have a range of experience in positions of increasing responsibility. Managers and shift supervisors should have a demonstrated leadership ability in relation to personnel (and, as appropriate, contractors) for whose activities they might become responsible.

3.27. Managers and supervisors should have a breadth of experience. Supervisors should have knowledge of the activities of all positions under their control. In higher managerial positions, the decisions that have to be taken will frequently involve knowledge not only of subordinate positions but also of matters outside the plant organization, such as company policies.

3.28. Plant managers should have experience in several key areas of the operation of the plant, such as operations, maintenance and technical support. This experience is usually gained over a period of at least 10–15 years, but of not less than five years. Plant managers should also have appropriate management experience.

3.29. Heads of operations, maintenance, quality assurance, training and technical support should have sufficient experience in their respective areas to develop specific competence and management ability. The head of operations should, in addition, have experience in reactor operations. This experience is usually gained over a period of at least five to eight years, with a minimum of two to three years at nuclear power plants, of which at least six months should be at the site concerned (or at a similar site).

3.30. The heads of nuclear safety and radiation protection and the persons responsible for reactor physics should have gathered specific experience at comparable facilities. Sufficient experience is usually gained over a period of at least four to six years at nuclear facilities, with a minimum of two to three years at nuclear power plants, of which at least six months should be at the site concerned (or at a similar site).

3.31. Shift supervisors should have experience in reactor operations at a nuclear power plant, in terms of both leading and working with a shift team. This experience is usually gained over a period of at least four to six years, with a minimum of two to three years at an operating plant, of which at least one year should be at the site concerned (or at a similar site).

Work experience for operators

3.32. Control room operators should have experience of working on shifts at nuclear power plants or at conventional power plants. Sufficient experience is gained over a period of at least three to four years, of which a minimum of two years should be at a nuclear power plant, with at least six months at the site concerned (or at a similar site).

3.33. All other operators should have acquired experience appropriate to their duties and responsibilities. In general, one year of experience should be considered a minimum for field operators [14].

Technicians and personnel with specific manual skills

3.34. Senior technicians and personnel with specific manual skills should have at least two to three years of practical experience. Other technicians and personnel with specific manual skills should have the appropriate experience to demonstrate the skills necessary to perform their duties and meet their responsibilities.

COMPETENCE AND QUALIFICATION OF CONTRACTORS AND SUPPLIERS

3.35. Paragraph 4.36 of GSR Part 2 [3] states that "The organization shall make arrangements for ensuring that suppliers of items, products and services important to safety adhere to safety requirements and meet the organization's expectations of safe conduct in their delivery."

All suppliers and contractors involved in design, engineering, manufacturing, construction, operation, maintenance or other safety related activities should be aware of the applicable safety requirements and expectations of the operating organization. Suppliers and contractors should also understand the safety culture of the operating organization (see Requirement 12 of GSR Part 2 [3]). This understanding is mutually beneficial for suppliers, contractors and the operating organization.

3.36. The operating organization should ensure that, during the entire period of the contracted work, contractor personnel involved in safety related activities are competent, qualified and medically fit to perform their assigned tasks.

3.37. The contractors selected for specific safety related activities should provide documentary evidence to the operating organization that they and their personnel have appropriate training and qualification (and, if necessary, the necessary certification) to perform the assigned work. This information should be provided before contractor personnel start such work, and should include confirmation of relevant experience in performing similar work.

4. THE APPROACH TO TRAINING NUCLEAR POWER PLANT PERSONNEL

4.1. Paragraph 4.20 of SSR-2/2 (Rev. 1) states:

> "Performance based programmes for initial and continuing training shall be developed and put in place for each major group of personnel (including, if necessary, external support organizations, including contractors). The content of each programme shall be based on a systematic approach. Training programmes shall promote attitudes that help to ensure that safety issues receive the attention that they warrant."

4.2. The operating organization should formulate an overall training policy. This policy should describe the commitment of the operating organization and managers to the training of personnel, and acknowledge the essential role of training in the safe and reliable operation and maintenance of the plant.

4.3. The training policy should be known, understood and supported by all relevant personnel. Managers, including the training manager, should be involved in developing the training policy.

4.4. A training plan should be prepared on the basis of the long term needs and goals of the plant. This plan should be reviewed periodically in order to ensure that it is consistent with current (and future) needs and goals. Factors that should be taken into account in the review of the training plan include: feedback of operating experience (see Requirement 24 of SSR-2/2 (Rev. 1) [1]); significant modifications to the plant or to the operating organization; changes in regulatory requirements; changes in the national education system; fluctuations in staffing; and specific staffing problems (e.g. loss of staff, lack of motivation) for plants that are approaching shutdown.

4.5. The training needs associated with the performance of safety related activities (see Requirement 8 of SSR-2/2 (Rev. 1) [1]) should be considered a priority, and relevant safety criteria, operating procedures, codes and standards, references and resources, tools and equipment should be used in the training. For such activities, training should include practical elements that are as representative of the actual job environment as possible.

4.6. Training should be used to foster and sustain a strong safety culture in accordance with Requirement 12 of GSR Part 2 [3]. Training should be fully encouraged and supported by managers, who should also be trained in fostering a strong safety culture.

4.7. Job specific training programmes should also be designed to develop skills and attitudes that contribute to safety.

4.8. For each position that performs safety related activities, the initial training needs and the continuing training needs should be established. These needs will vary depending on the individual position, the level of responsibility and the level of competence, and should be determined by persons with specific competence in plant operation and experience in developing training activities. These training needs should relate to the tasks and activities to be performed and include a clear focus on safety.

4.9. The operating organization should ensure the following with regard to personnel performing safety related activities:

(a) Training needs are continuously analysed, in accordance with para. 4.18 of SSR-2/2 (Rev. 1) [1], and this analysis gives priority to safety.
(b) A training programme is developed, in accordance with para. 4.19 of SSR-2/2 (Rev. 1) [1].
(c) All necessary resources and facilities for implementing the training programme are provided.
(d) The performance of trainees is assessed at various stages of the training.
(e) The effectiveness of the training is evaluated, in accordance with para. 4.23 of GSR Part 2 [3].
(f) The competence of personnel is periodically checked, and continuing training or retraining is provided on a regular basis, in accordance with para. 4.19 of SSR-2/2 (Rev. 1) [1].

4.10. The operating organization should establish a training entity that is responsible for assisting the plant manager in establishing, verifying and maintaining the competence of personnel. Even if off-site training facilities are to be used, a training entity should still be included in the plant organization. The training entity should advise the plant manager on all matters relating to training, coordinate training activities on the site, ensure proper liaison with off-site training facilities, and collect records of the satisfactory completion of initial training and continuing training of individuals.

4.11. Paragraph 4.18 of SSR-2/2 (Rev. 1) [1] states:

> "The management of the operating organization shall be responsible for the qualification and the competence of plant staff. Managers shall participate in determining the needs for training and in ensuring that operating experience is taken into account in the training. Managers and supervisors shall ensure that production needs do not unduly interfere with the conduct of the training programme."

The existence of a training entity should not relieve line managers of their responsibility to ensure that their personnel are adequately trained and qualified. Supervisors should recognize and make provision for the training needs of their subordinates. The responsibilities and authority of training personnel, as distinct from those of line managers, should be clearly defined and understood.

4.12. Consideration should be given to enhancing training programmes for personnel at older plants to compensate for losses of personnel due to retirement or job changes. Training programmes should also be adapted to accommodate any special technical, administrative and operational needs of older plants.

4.13. Owing to the trend towards the automation of plants, operators might need to interpret greater amounts of plant information, and more complex plant equipment might need to be maintained. Training programmes should reflect these changes; for example, there might need to be a greater focus on structured fault finding and decision making.

SYSTEMATIC APPROACH TO TRAINING

4.14. A systematic approach to training should be used for personnel [15–17]. The systematic approach provides a logical progression, from identification of the competences necessary for performing a job, to the development and implementation of training towards achieving these competences, and to the subsequent evaluation of this training.

4.15. A systematic approach to training includes the following phases:

(a) Analysis. This should comprise the identification of training needs and of the competences necessary to perform a particular job.
(b) Design. In this phase, competences should be converted into training objectives. These objectives should be organized into a training plan.
(c) Development. In this phase, training materials should be prepared so that the training objectives can be achieved.
(d) Implementation. In this phase, training should be conducted using the training materials developed.
(e) Evaluation. In this phase, all aspects of the training programmes should be evaluated on the basis of data collected in each of the other phases (e.g. operating experience data, performance indicators, modification data, procedure changes and inputs from supervisors and job incumbents). This should be followed by feedback leading to improvements in the training programmes and to plant improvements.

TRAINING SETTINGS AND METHODS

4.16. Training should be carefully controlled and structured to achieve the training objectives in a timely and efficient manner. The following training settings and methods should be considered:

(a) The classroom is the most frequently used training setting. Its effectiveness should be enhanced by the use of appropriate training methods such as lectures, discussions, role playing, critiquing and briefing. Training aids and materials such as written materials, presentations, audio and video based materials, scale models and simulators should be used to support classroom instruction, as necessary.
(b) On the job training should be conducted in accordance with guidelines developed by experienced personnel who have been trained to deliver this form of training. Progress should be reviewed, and assessments should be performed by an independent assessor.
(c) Simulator based training for control room operators, shift supervisors, responsible managers and technical support personnel should be conducted. The simulator should be equipped with software of sufficient scope to cover normal operation, anticipated operational occurrences and a range of accident conditions. Other personnel may also benefit from simulator based training.

(d) Training mock-ups and models should be provided for activities that need to be performed quickly and skilfully and which cannot be practised with actual equipment. Training mock-ups should be full scale if practicable.
(e) Training should be provided in laboratories and workshops to ensure safe working practices in these environments.
(f) Self-study training should be encouraged. This does not have to be undertaken at a training facility, but in all cases the trainees should have support from a designated expert.

Typically, training should consist of periods of formal training in the classroom mixed with intervals of simulator, laboratory or workshop training, and should include practical training at the plant.

4.17. Plant commissioning provides an important opportunity for hands-on training for operating personnel and personnel in supporting functions. For example, before fuel is loaded at a new plant, the testing of certain components and systems can be undertaken with a freedom of access that is not possible later in the plant's operating lifetime.

4.18. The training of control room operators should include classroom training, on the job training and simulator training (see para. 4.19). Simulator sessions should be structured and planned in detail to ensure adequate coverage of the training objectives and to avoid any limitations associated with the simulation. The sessions should include preliminary briefings and follow-up reviews.

4.19. Representative simulator facilities are required to be available (see para. 4.24 of SSR-2/2 (Rev. 1) [1]), and these should be used for the training of control room operators, shift supervisors, responsible managers and technical support personnel. With regard to simulator training, consideration should be given to the following:

(a) Administrative procedures should be developed for the design, development, implementation and evaluation of all training conducted on the simulator.
(b) Line managers should be involved in the identification of simulator training needs and in observing and assessing simulator exercises.
(c) Simulator scenarios should be carefully prepared, including objectives and criteria for termination.
(d) Simulator exercise guides for the conduct of simulator training should be developed for all demonstration, training and evaluation scenarios.
(e) Simulator training exercises should be planned systematically and performed at a frequency that reflects the needs of the personnel being trained.

(f) Training should be conducted using a shift team concept to develop team skills, good communication and coordination habits and trust in the application of plant procedures.
(g) Individual and team assessments should be based on predetermined performance criteria.
(h) Simulator instructors should be selected based on both human and technical competences and receive initial and continuing instructor training (see also para. 4.23 of SSR-2/2 (Rev. 1) [1]).

4.20. The importance of training by means of simulators and computer based systems should be emphasized in order to develop human–machine interface skills.

4.21. All progress made in training should be assessed and documented. The means of assessing a trainee's ability include written examinations, oral questioning and performance demonstrations. A combination of written and oral examinations has been found to be the most appropriate form of demonstrating knowledge and skills. In the assessment of simulator training, predesigned and validated observation forms and checklists should be utilized in order to increase objectivity. All assessments of simulator training sessions should include an evaluation of the trainees and of the feedback given, and further measures should be considered as a result of the evaluation. Assessment should not be regarded as a one-off activity. Reassessment of individuals by instructors and their immediate supervisors should be undertaken at regular intervals.

INITIAL AND CONTINUING TRAINING

4.22. The training programme is required to include initial training and continuing training or retraining (see para. 4.19 of SSR-2/2 (Rev. 1) [1]). Initial training should be provided to individuals before they are assigned to a job or a position within the operating organization. Continuing training should be provided for all personnel throughout their working life to ensure that they maintain the necessary knowledge, skills and attitudes. Continuing training should also focus on improving the skills and attitudes that are necessary for safety related activities. Retraining should be provided in cases (e.g. extended absence from duties) where it has not been possible to provide continuing training. Retraining also describes training in a different knowledge, skill or attitude, for example, because of a major modification to the plant or to plant operation, the installation of a new plant or a change of duties. The training programme for each individual should define the contents of the initial training, and the continuing training or retraining, as appropriate. Special training might be necessary if an individual shows deficiencies in performance or if there is a need to prepare for a non-routine activity or event.

4.23. Training for all personnel of the operating organization, including plant personnel, should include general induction training (see paras 5.1 and 5.2) as well as specific training to ensure they have a thorough understanding of their particular duties and responsibilities and of their contribution to the safe and efficient operation of the plant.

4.24. An initial training programme should be established to ensure that all personnel achieve the necessary competence to perform their assigned duties. Initial training should help personnel to achieve a high level of performance in terms of safety and professionalism. The goals of initial training should include the following:

(a) To complement any formal education in the areas of technology and science;
(b) To provide knowledge and understanding of nuclear safety principles, for example, defence in depth;
(c) To provide an understanding of safety management, procedures and expected levels of performance;
(d) To provide knowledge of nuclear technology and of the plant, as relevant to the assigned duties;
(e) To provide an understanding of the principles of operation and maintenance of specific plant systems and equipment;
(f) To develop specific skills relating to the assigned duties;
(g) To emphasize general safety aspects of the plant, and specific safety aspects relating to the assigned duties;
(h) To encourage an appropriate attitude towards safety.

4.25. The goal of continuing training is to maintain a high level of performance of personnel. To achieve this goal, areas of knowledge necessary for safe plant operation should be systematically reviewed. The training programme is required to incorporate operating experience from the plant and from industry (see para. 4.22 of SSR-2/2 (Rev. 1) [1]). Continuing training should also address identified problems in performance, plant modifications and procedural changes. The aims of continuing training should include the following:

(a) To improve the knowledge and skills of personnel when changes in their assigned duties are identified;

(b) To maintain (and, in selected areas, enhance) the skills and knowledge necessary to perform their assigned duties in different operational states and in accident conditions, in accordance with para. 4.17 of SSR-2/2 (Rev. 1) [1];
(c) To increase the level of understanding of issues such as nuclear safety principles and safety requirements that were presented in initial training, with emphasis on areas of demonstrated weakness;
(d) To maintain an awareness of the responsibility for safe operation of the plant and of the consequences of shortcomings in human performance;
(e) To correct deficiencies in personnel performance that have been detected through the analysis of plant operating experience;
(f) To maintain knowledge of plant modifications and understanding of procedural changes in areas to which personnel are assigned;
(g) To emphasize lessons from industry and plant specific operating experience;
(h) To emphasize topics identified by managers and supervisors;
(i) To enhance the performance of operating personnel through timely training for infrequent, difficult and important tasks.

4.26. As part of continuing training, personnel should be periodically reminded of concepts in areas such as reactor physics, principles of operation of plant systems and equipment, thermohydraulics, plant chemistry, reactor safety, non-radiation-related safety and radiation protection.

4.27. Continuing training or retraining might also include training to improve the career development potential of selected individuals.

4.28. Continuing training should be provided for operators at appropriate intervals to ensure that the necessary knowledge and skills for the safe and efficient operation of the plant are retained and refreshed.

4.29. Structured continuing training or retraining for control room operators, shift supervisors, responsible managers and technical support personnel should be given on a representative simulator. Simulator training exercises should be performed annually. Such exercises should reflect operating experience with emphasis on those situations that do not occur frequently, for example, startup, shutdown, special transients, accident conditions, including during shutdown mode. Teamwork should be emphasized in dealing with incidents and accidents.

4.30. The time necessary for all personnel to undergo formal continuing training on a regular basis should be taken into account when work schedules are established. For maintenance personnel, refresher training should be given on maintenance activities that are performed infrequently.

TRAINING FOR EMERGENCIES AND ACCIDENT MANAGEMENT

4.31. Requirements for emergency preparedness and response — including training, drills and exercises — are established in IAEA Safety Standards Series No. GSR Part 7, Preparedness and Response for a Nuclear or Radiological Emergency [18].

4.32. Paragraph 5.5 of SSR-2/2 (Rev. 1) [1] states:

> "A training programme for emergencies shall be established and implemented to ensure that plant staff and, as required, staff from other participating organizations possess the essential knowledge, skills and attitudes required for the accomplishment of non-routine tasks under stressful emergency conditions."

4.33. Paragraph 5.6 of SSR-2/2 (Rev. 1) [1] states:

> "The emergency plan shall be tested and validated in exercises before the commencement of fuel loading. Emergency preparedness training, exercises and drills shall be planned and conducted at suitable intervals, to evaluate the preparedness of plant staff and staff from external response organizations to perform their tasks, and to evaluate their cooperation in coping with an emergency and in improving the efficiency of the response".

4.34. The purpose of exercises and drills should be as follows:

(a) To demonstrate how effectively the emergency plan (see Requirement 18 of SSR-2/2 (Rev. 1) [1]), or part of the emergency plan, can be implemented;
(b) To confirm the adequacy of the plan to deal with an emergency and to identify potential improvements;
(c) To verify that the appropriate lines of communication are established and maintained;
(d) To verify that all participating individuals are familiar with, and capable of performing, the emergency duties assigned to them;
(e) To verify that emergency response duties and all related duties can be performed under stressful conditions in a timely manner and in accordance with the expected schedule;
(f) To verify the provisions for the assessment of radiological hazards and the implementation of protective actions.

4.35. Training should be provided for all personnel at the plant who have assigned duties in the emergency plan. The training for emergencies should include non-radiation-related safety, in particular firefighting and medical first aid. Training should also be provided for on-site staff who have no specific emergency duties, to familiarize them with the procedures for alerting personnel to emergency conditions. Similar training, or at a minimum a well-structured information briefing, should be provided to contractor personnel or other temporary personnel.

4.36. Full scale exercises involving external organizations (e.g. the police, fire services, ambulance teams, rescue teams and other emergency services) are required to be arranged (see para. 6.30 of GSR Part 7 [18]). The exercises should include extended aspects, such as the need for corporate and national level coordinated arrangements, the response to events of long duration and to events involving several units on a site simultaneously. Exercise scenarios should be carefully prepared, including objectives and criteria for termination. The conduct of a plant exercise should not create any conditions that could jeopardize plant safety. Further information on the conduct of exercises is provided in Ref. [19].

4.37. Training is also required to be provided on accident management, in accordance with para. 5.8E of SSR-2/2 (Rev. 1) [1]. This training is required to address accidents more severe than design basis accidents and should provide the knowledge and skills to effectively manage such accidents. Further recommendations on training, exercises and drills for accident management are provided in IAEA Safety Standards Series No. SSG-54, Accident Management Programmes for Nuclear Power Plants [20].

4.38. Specific training should be provided for those personnel who perform specialized duties in the event of an accident. For example, topics such as nuclear safety analysis, compliance with regulatory requirements, the application of relevant codes and standards, the evaluation of safety margins for the plant and the application of symptom oriented procedures should be covered. The main results of any probabilistic safety assessment, showing the importance of plant systems in preventing damage or severe accidents, should be included in the training programme.

4.39. Specific in depth training in emergency operating procedures should be provided to all relevant operating personnel. One of the aims of this training is to prevent any degradation in human performance that might occur in stressful situations. Classroom training should be used to explain the conceptual basis, terminology and structure of emergency operating procedures, and the roles and responsibilities of individuals in the implementation of these procedures. The implementation of emergency operating procedures should be practised using simulator based training, to provide operating personnel with the necessary knowledge and skills.

4.40. Training on the implementation of emergency operating procedures should include the following:

(a) A description of how the plant reacts to various types of initiating event, using appropriate graphics. The description should be based on best estimate calculations or on actual operating experience. Alternatives for each type of event should be presented to show how the plant can be returned to a safe state.
(b) A discussion of the basic accident management strategy for each type of event, and possible alternatives. Results from the analysis of each of these strategies should be presented.
(c) An explanation of the principles of ensuring plant safety by ensuring the fulfilment of fundamental safety functions (see Requirement 4 of SSR-2/1 (Rev. 1) [2]).
(d) An explanation of the logic and organization of the emergency operating procedures, including the roles of individual operating personnel.
(e) A description of the methods to be used to return the plant to a safe state, and a discussion of the purpose of each step, or group of related steps, in the emergency operating procedures.
(f) An explanation of the criteria under which emergency operating procedures could be modified.

4.41. Specific training should be provided on the implementation of severe accident management guidelines (see SSG-54 [20]). This training may be conducted in a combination of settings, including simulation, emergency drills and classroom training. Control room simulators are usually not validated for severe accident conditions, and a high degree of caution should be applied to their use in the training of operating personnel for such conditions. However, simulators can be used in exercises for initial accident classification and decision making. Workstations or other computer systems that simulate the evolution of accidents after core damage has occurred should be used to the extent practicable.

4.42. Managers and supervisors should be trained in directing operating personnel (using available information, plant systems and equipment) to mitigate the consequences of severe accidents. Operating personnel should be trained in recognizing situations in which the emergency operating procedures are not adequate and the severe accident management guidelines should be used. The transition from emergency operating procedures to severe accident management guidelines should be part of this training. Exercises should be designed to ensure that the decision making function is tested and is clearly understood by the accident management team.

4.43. Training of the managers and the technical specialists involved in accident management should include the following:

(a) Diagnosing the causes of the accident and assessing the potential effects:
   (i) Assessing the status of the core, the containment and of important safety systems;
   (ii) Predicting the probable timing of key events in the evolution of the accident;
   (iii) Assessing core damage;
   (iv) Anticipating problems likely to cause the situation to deteriorate further;
   (v) Estimating the possible pressure rise, and temperature rise from hydrogen combustion or reactor vessel failure.
(b) Formulating the strategy for accident management:
   (i) Identifying and assessing accident management strategies to prevent or halt core damage, to prevent containment failure and to prevent or reduce radioactive releases;
   (ii) Using available information, including evaluations from probabilistic safety assessment, to set priorities for corrective actions.
(c) Taking corrective actions:
   (i) Taking actions to re-establish the redundancy, diversity and independence of safety systems, and integrating these actions with those of control room operators;
   (ii) Implementing actions to halt core damage, prevent containment failure and to prevent or reduce radioactive releases.
(d) Monitoring and updating the accident management strategy:
   (i) Monitoring the effectiveness of actions implemented by control room operators;
   (ii) Anticipating problems that might further degrade the core and the safety systems.

4.44. The training programme for accident management should be reviewed periodically and updated, as necessary, to take into account new knowledge and operating experience.

TRAINING DOCUMENTATION

4.45. Training documentation should consist of records, reports and feedback associated with the training programmes and with the performance of trainees. The documentation should be used to assist managers in monitoring the effectiveness of the training programme (see para. 4.23 of GSR Part 2 [3]), as well as in an annual management review of the competence of personnel. The documentation should also provide a historical record of the changes made to the training programme as a result of evaluation and feedback.

4.46. The operating organization should maintain adequate documentation of the training of individuals (including on the job training) and of the performance of individual trainers and trainees (including a list of main activities performed). The documentation should include (or at least provide a reference to) learning objectives, lesson and exercise plans, student reading material, guidance for on the job training, and documentation on instructors and assessors. The aims of this documentation should be as follows:

(a) To provide evidence of the competence of all personnel whose duties might affect the safety of the plant;
(b) To provide evidence of the authorizations issued for certain operating positions (see para. 4.16 of SSR-2/2 (Rev. 1) [1]);
(c) To enable managers to deploy personnel effectively, ensuring that only suitably qualified and experienced personnel are assigned to safety related activities;
(d) To provide the information necessary for reviews of the training programme and for corrective actions, if necessary;
(e) To provide the documentation necessary to meet regulatory requirements (e.g. with regard to the granting or renewing of authorizations for the plant).

4.47. Documentation from training programmes should be retained to enable the review of the training contents, schedules and the results of current and past programmes. This documentation should be classified in accordance with the type of document and the necessary retention period, and should be appropriately stored, organized and indexed for ease of retrieval.

4.48. The administration, storage and safe keeping of training documentation should be in accordance with Requirement 15 of SSR-2/2 (Rev. 1) [1] and with the management system established in accordance with Requirements 6–8 of GSR Part 2 [3] and Requirement 2 of SSR-2/2 (Rev. 1) [1].

4.49. The training entity should report periodically to appropriate levels of management on the status and effectiveness of training activities. Significant events or problems in the implementation of the training programme should be identified and reported in a timely manner.

5. TRAINING PROGRAMMES FOR NUCLEAR POWER PLANT PERSONNEL

5.1. All new personnel starting work at a plant should be inducted into the organization and their working environment in a systematic and consistent manner. General personnel training programmes should give new personnel a basic understanding of their responsibilities and of safe and secure work practices, the importance of quality management and of following procedures, and the practical means of protecting themselves from the hazards associated with their work. The amount of training to be provided on specific topics should be commensurate with the assigned duties of personnel. The basic principles of safety culture (see Ref. [21]) should be taught to all personnel, and refresher training on general topics should also be provided periodically.

5.2. General induction training should be provided to each member of personnel or contractor working at the plant, to address the following:

(a) Introduction to the plant organization and administration;
(b) Nuclear safety principles (e.g. defence in depth);
(c) The management system;
(d) Safety culture;
(e) Non-radiation-related safety (e.g. electrical safety, rigging and lifting, work in confined spaces, chemical hazards, use of protective equipment, first aid);
(f) Radiation protection, including techniques for the optimization of radiation protection;
(g) Foreign material exclusion (see SSG-76 [9]);
(h) Fire protection, including fire prevention;
(i) Environmental protection;
(j) Use of human performance tools;
(k) Nuclear security and access control;
(l) Emergency alarms, escape routes and assembly points.

5.3. Training programmes for most positions at a nuclear power plant should include on the job training, to ensure that trainees obtain the necessary knowledge and skills in their actual working environment. Formal on the job training provides hands-on experience and allows the trainee to become familiar with plant routines. However, on the job training does not simply mean working under the supervision of a qualified individual; it also involves the use of training objectives, qualification guidelines and trainee assessments. This training should be conducted and evaluated in the working environment by qualified, designated individuals.

5.4. Training programmes should include training in any new technologies and equipment that are introduced at the plant.

5.5. Suitable personnel should be trained in root cause analysis and the assessment of human and organizational factors, with the aim of creating a pool of staff who can evaluate events objectively and make recommendations on how to avoid their recurrence.

5.6. All training programmes for specific plant activities should make reference to the need to sustain a strong safety culture (see paras 3.3, 3.23 and 4.6). These programmes should stress the need for an understanding of safety issues, should include consideration of the possible consequences of shortcomings in human performance and should deal specifically with ways in which such shortcomings can be avoided or corrected.

5.7. All personnel who are likely to be occupationally exposed to ionizing radiation should receive suitable training to understand radiation risks and the technical and administrative means of optimizing protection and safety in accordance with Requirement 11 of IAEA Safety Standards Series No. GSR Part 3, Radiation Protection and Safety of Radiation Sources: International Basic Safety Standards [22].

5.8. Training programmes for positions such as managers, nuclear safety experts and technical specialists, control room operators and senior technicians should provide a thorough understanding of the basic principles of nuclear technology, nuclear safety and radiation protection, and the design intents and assumptions, together with the necessary on the job training. The training programme for other personnel (including technicians and personnel with specific manual skills) should be more practical, with supporting explanations of the underlying theory and of safety related aspects.

5.9. Training programmes should address the interface between safety and nuclear security, to ensure that personnel understand how this interface affects their assigned duties.

5.10. Training programmes should include training for personnel that might need to work in extremely stressful situations, in order to increase the ability of the operating organization to cope with such situations. Stressful situations that could impact decision making and reduce personnel effectiveness (e.g. during a natural disaster or a nuclear or radiological emergency) should be considered. Training on the effects of psychological stress in difficult work environments (e.g. noisy, unlit, smoke-filled) or following a significant event at the plant should be provided for personnel who might experience such conditions in performing their duties.

5.11. Training programmes should address safety during outages. The training should include measures such as effective planning of work, work management, safety assessment, management of plant configuration, testing after maintenance and modifications, system alignments, review of experience and accident management. Further recommendations on outage management are provided in SSG-74 [8].

5.12. Relevant personnel specified by the operating organization should be made familiar with the features of deterministic safety assessment, probabilistic safety assessment and risk informed applications as part of their training programme.

5.13. Training programmes should include training for emergencies and accident management (see paras 4.31–4.44). If the simulator facilities are not validated for severe accidents (see para. 4.41), computer based training, classroom training and plant walkthroughs should be used to explain the consequences of an accident involving significant core degradation.

5.14. Training programmes should include training on any modifications to the plant, to ensure that relevant personnel are familiar with the modifications and have the necessary knowledge and skills to operate and maintain modified equipment in a safe and reliable manner. This training should be completed before the commissioning, operation and maintenance of modified equipment. Training programmes should also include training on modifications to the structure of the operating organization or to processes within the organization. Further recommendations on plant modifications are provided in SSG-71 [5].

5.15. Recommendations on training programmes for different groups of personnel are provided in paras 5.16–5.41.

TRAINING PROGRAMMES FOR MANAGERS AND SUPERVISORS AT NUCLEAR POWER PLANTS

5.16. Training programmes for managers and supervisors should emphasize the need to foster a strong safety culture (see Requirement 12 of GSR Part 2 [3]), and should include training in making successful presentations of safety related messages to personnel. This training should assist managers and supervisors in promoting an awareness that safety should be considered a primary objective in their day-to-day activities, and be given priority over production needs.

5.17. The training of managers and supervisors should emphasize the special features of managing a nuclear power plant, with its primary focus on safety, and the need for familiarity with emergency procedures. Special attention should be given to the importance of maintaining a high level of protection and safety and to gaining the benefits of feedback of operating experience and root cause analysis of events that occur at the plant.

5.18. The training programmes for managers and supervisors should ensure that they have a thorough understanding of all the relevant standards, rules and regulations, as well as a good overall knowledge of the plant and its systems. Individuals who contribute to the management or supervision of emergency response should be specially trained for these duties.

5.19. Training programmes for managers and supervisors should include training on managerial and supervisory skills, coaching and mentoring, self-assessment techniques, root cause analysis, human and organizational factors, team building, communication, management system reviews (see Requirement 13 of GSR Part 2 [3]) and other safety reviews periodically performed at the plant.

5.20. The career development of managers should include involvement with external groups, networks and bodies at the national and international level.

TRAINING PROGRAMMES FOR OPERATING PERSONNEL

5.21. The main objectives of the training of control room operators should be to develop and maintain sufficient knowledge and skills to ensure that they are able to perform the following actions from the main control room and, when applicable, from the supplementary control room:

(a) Monitor and control the status of plant systems in accordance with operating procedures and operational limits and conditions (see SSG-70 [4]);
(b) Conduct operations in a safe and reliable manner, without causing excessive thermal or mechanical load to plant equipment;
(c) Take correct actions in response to various abnormal conditions, and bring the plant to a safe state, including shutdown, whenever needed;
(d) Take actions to prevent accidents, including severe accidents, and to mitigate the consequences of accidents if they do occur.

5.22. The training of operating personnel should cover relevant areas of technology to a level of detail that is commensurate with their assigned tasks. It should include a thorough knowledge (theoretical and practical) of plant systems and their functions, and of the plant layout and operation. Participation in the stages before the operation of a new plant is a valuable opportunity for such training. Emphasis should be placed on items that are important to safety.

5.23. The results of the probabilistic safety assessment of the plant should be used in the training of operating personnel to demonstrate the importance of plant systems in preventing plant damage or severe accidents. The training should emphasize the importance of maintaining the plant within operational limits and conditions, and the consequences of not complying with these limits. The importance of maintaining reactivity control and core cooling at all times, including the period when the plant is not in operation, should also be emphasized in the training.

5.24. Control room operators should also be trained in plant diagnostics, control actions and administrative tasks and human factors such as attitudes and human–machine and human–human (teamwork) interfaces.

5.25. Operating personnel should be trained to be aware of the locations of all significant amounts of radioactive material in the plant, and of the controls to be applied to them.

5.26. Operating personnel should be trained in routines for normal operation of the plant and in the response of the plant to changes that could cause accidents if not counteracted. The training programme should aim to improve the diagnostic skills of the trainees. Operating procedures for normal operation, anticipated operational occurrences and, as far as practicable, accident conditions should be included in the training programme and should be practised using the simulator, so that the trainees recognize the consequences of any shortcomings in the implementation of these procedures.

5.27. Field operators should receive training commensurate with their duties and responsibilities. The main objectives of this training should be to develop and maintain adequate knowledge and skills to ensure that they are able to perform the following tasks:

(a) Monitoring the performance and status of plant equipment, and recognizing any deviations from normal conditions, including any non-compliance with the foreign material exclusion programme;
(b) Conducting operations in a safe and reliable manner, without causing unacceptable risks to the plant;
(c) Detecting and properly responding to plant conditions with the goal of preventing or, at a minimum, of mitigating unanticipated plant transients;
(d) Implementing the emergency operating procedures and severe accident management guidelines outside the main control room.

All personnel in this category should have detailed knowledge of the operational features of the plant as a whole, and hands-on experience.

5.28. Reactor shutdown or low power operating states contribute significantly to the risk of core degradation; consequently, ensuring safety during plant maintenance, plant modification, low power operating states and shutdown should be emphasized in the training of operating personnel. These activities sometimes involve unusual plant configurations and make extra demands on the knowledge and skills of the operating personnel. Suitable training should be provided in advance of these activities to reduce the risks to the plant and to operating personnel.

TRAINING PROGRAMMES FOR MAINTENANCE PERSONNEL

5.29. Training programmes for maintenance personnel should emphasize the potential consequences for safety relating to any technical or procedural issues associated with maintenance activities. Experience of faults and hazards caused by maintenance activities at the plant, or at other plants and in other industries, as appropriate, should be reviewed and taken into account in training programmes. An appropriate emphasis should be placed on safety culture in all aspects of the training for maintenance personnel.

5.30. Training programmes for maintenance personnel should include training on the plant layout and the general features and purposes of plant systems as well as quality management. Training should include all relevant maintenance activities and procedures, including testing, surveillance and inspection. Fault finding and special maintenance skills (e.g. avoidance and detection of human induced common cause failure, foreign material exclusion) should also be covered.

5.31. Training programmes for maintenance personnel may include training given by component manufacturers, training on equipment mock-ups or on the job training under the supervision of experienced staff. Maintenance personnel should have access to mock-ups and models for training in maintenance activities that have to be completed quickly and that cannot be practised with actual equipment.

5.32. Past incidents relating to shortcomings in the performance of maintenance activities should be recreated in the training using a mock-up that can reproduce complex situations (e.g. involving difficulties with techniques, access or radiation exposure), so that the lessons from these incidents can be demonstrated, and the capabilities of maintenance personnel in these situations can be evaluated.

5.33. The concept of 'just in time' training can be used as part of the training programme for maintenance personnel. Specific task oriented training should be included in the work schedule and should be provided shortly before the task is performed.

TRAINING PROGRAMMES FOR OTHER TECHNICAL PERSONNEL

5.34. Personnel involved in plant chemistry, radiation protection, nuclear engineering, fuel handling, quality management or other technical functions should undergo individual qualification and receive training as appropriate to their assigned duties.

5.35. Technicians might be assigned to perform work similar to their own at other plants or with equipment suppliers. Emphasis should be placed on the development of specific practical skills, with the minimum of classroom training that is considered necessary. In some cases, laboratories or simulators may be provided by equipment suppliers, particularly suppliers of items important to safety.

5.36. Personnel with specific manual skills should undergo general training (see paras 5.1 and 5.2) and specific training (see paras 5.3–5.14) to develop the basic and specific skills necessary for their assigned duties. The training programme could include placing such personnel on secondment with suppliers of equipment and components and with construction groups. Certain skills may also be developed with the help of mock-ups.

TRAINING PROGRAMMES FOR TRAINING INSTRUCTORS

5.37. Paragraph 4.23 of SSR-2/2 (Rev. 1) [1] states:

> "All training positions shall be held by adequately qualified and experienced persons, who provide the requisite technical knowledge and skills and have credibility with the trainees. Instructors shall be technically competent in their assigned areas of responsibility, shall have the necessary instructional skills, and shall also be familiar with routines and work practices at the workplace. Qualification requirements shall be established for the training instructors."

5.38. Training instructors should thoroughly understand all aspects of the contents of the training programmes and the relationship between these contents and plant operation. In addition, the instructors should be familiar with methods of adult learning and a systematic approach to training, and should have the necessary skills to assess the progress made by trainees.

5.39. All personnel in the training entity, including simulator and technical support personnel, should be given training commensurate with their assigned duties. Instructors should be allowed the time necessary to maintain their technical and instructional competence, by secondment to an operating plant on a regular basis and by continuing training.

5.40. Personnel in the on-site training entity should receive training on the policies of the operating organization, in particular with regard to safety management and safety culture, compliance with regulatory requirements and quality management.

5.41. Training provided by external organizations should be evaluated to ensure that it is of sufficient quality and meets the needs at the plant.

PERIODIC REVIEW OF TRAINING PROGRAMMES

5.42. Paragraph 4.21 of SSR-2/2 (Rev. 1) [1] states that "The training programmes shall be assessed and improved by means of periodic review." The review should cover the adequacy and effectiveness of training with respect to the actual performance of personnel in their jobs. The review should also examine training needs, the training plan and the training programmes, training facilities and materials, and training records, and should consider whether these adequately address any changes to regulatory requirements, modifications to the facility and recent feedback from operating experience. The review of the training of personnel is an integral component of the systematic approach to training (see para. 4.15(e)).

5.43. The review of training programmes should be undertaken by persons other than those directly responsible for the training. Plant managers should be directly involved in the evaluation of training programmes. Close cooperation should be maintained in the training evaluation process between plant managers, individual departments and the training entity.

5.44. In the review of training programmes, the following sources of information on the effectiveness of training programmes and on factors influencing training needs should be considered:

(a) Feedback from managers and supervisors, trainees and instructors (see para. 5.49) and other personnel.
(b) Operating experience.
(c) Information from events at the plant or at other plants, including root cause analysis and corrective actions (see para. 5.48).
(d) Problems in the training process, including failure of trainees in the assessments.
(e) Deficiencies in the performance of personnel. The conduct of operations and maintenance activities, including compliance with requirements for radiation protection and for non-radiation-related safety, should be monitored to identify any problems due to incorrect or insufficient training.
(f) Team issues relating to command, control and communication.

5.45. Training programmes should be reviewed, and training needs should be determined for any plant modifications or changes (see also para. 4.43 of SSR-2/2 (Rev. 1) [1]). Examples of such modifications include the following:

(a) New or modified plant equipment;
(b) New or revised procedures;
(c) New or updated software;
(d) New or modified regulatory requirements;
(e) New job expectations or training expectations;
(f) Modified organizational structure.

5.46. Training instructors should also regularly visit plants and work areas within plants to observe the performance of personnel, in order to improve their understanding of specific training needs.

5.47. A process should be established to routinely provide information to the training entity before proposed plant modifications or changes in plant procedures, so as to allow sufficient time for appropriate training to be provided. This is particularly important in relation to simulator training, for which the time needed to modify simulator hardware and software can be significant.

5.48. The operating organization should make every effort to analyse events in order to identify root causes, especially those relating to human factors. The results of such analyses should be fed back as appropriate into relevant training programmes. Plant event reports and industrial accident reports can identify tasks in which inadequate training might be contributing to equipment damage, excessive unavailability of equipment, unscheduled maintenance, a need for repetition of work, unsafe practices or a lack of adherence to approved procedures. This information should be supplemented by means of interviews with those concerned.

5.49. Trainees and training instructors can provide useful feedback for improving the training programmes. A questionnaire should be completed by trainees and trainers after each training element, and this should focus on the effectiveness of the training and on ways in which it could be improved.

5.50. As part of the review, an action plan to improve the training programmes should be developed and implemented. This may include improvements in the conduct of training and/or changes in the training programmes.

5.51. An independent review of the training plan for the plant (see para. 4.4) should be conducted by an external organization. This external review should be considered complementary to the internal periodic review of the training programmes. The results of the external review should be integrated with the results of the internal review, to identify necessary changes and improvements in the training programmes.

6. TRAINING FACILITIES AND MATERIALS

6.1. Paragraph 4.24 of SSR-2/2 (Rev. 1) [1] states:

> "Adequate training facilities, including a representative simulator, appropriate training materials, and facilities for technical training and maintenance training, shall be made available for the training of operating personnel. Simulator training shall incorporate training for plant operational states and for accident conditions."

6.2. The training facilities should provide for classroom training, computer based training, simulator training and individual studies. The training materials provided should help the trainees understand the plant and its systems. Detailed technical information to be used as reference material should also be available in the training facilities. The effectiveness of classroom training should be enhanced by the use of visual aids.

6.3. Consideration should be given to the use of computer based multimedia training packages and distance learning techniques (e.g. e-learning).

6.4. The simulator should include the following features:

(a) A replica of the main control room and the supplementary control room;
(b) A realistic working environment, including aspects such as the use of documentation, logging systems and communication systems;
(c) A behaviour that effectively simulates the behaviour of actual plant systems;
(d) The ability to simulate randomly selected failure combinations, severe transients, and infrequent and abnormal situations that have a low probability of occurrence;
(e) The ability to model auxiliary systems;
(f) Instructor aids (e.g. isolated booth, a means for freezing and reversing the simulation scenarios, automatic recording of the actions of trainees and the behaviour of systems, video cameras and recording devices).

6.5. Maintenance personnel and technical support personnel should have access to workshops, laboratories and facilities that are equipped with mock-ups, models and actual components that enable these personnel to be trained in activities that cannot be practised with installed equipment (e.g. because of high dose rates).

6.6. A procedure is required to be established for the periodic review and timely modification and updating of training facilities and materials, to ensure that they accurately reflect all modifications and changes made to the plant (see para. 4.21 of SSR-2/2 (Rev. 1) [1]).

7. AUTHORIZATION OF CERTAIN OPERATING POSITIONS

7.1. The formal authorization of certain operating positions, as referred to in para. 4.16 of SSR-2/2 (Rev. 1) [1], is the granting of written permission for an individual to conduct specified activities and to meet specified responsibilities.

7.2. The operating organization should establish a procedure by which persons controlling or supervising changes in the operational status of the plant, or who have other duties directly relating to safety, are authorized before they are allowed to perform their assigned duties. This procedure should include an assessment of the competence of persons to be authorized (see para. 7.5).

7.3. Work on structures, systems or components important to safety that is performed by contractor personnel should be formally authorized as described in para. 7.2. Such personnel should also be supervised by a representative of the operating organization who also meets the competence criteria established for such work.

7.4. At a minimum, the persons who occupy the following positions should be formally authorized:

(a) The shift personnel that directly supervise the operation of the plant (or unit) and who decide on safety measures during normal operation, anticipated operational occurrences and accident conditions. These personnel (normally the shift supervisor and the deputy shift supervisor) give commands to the other personnel in the shift and are responsible for the safe performance of the plant (or unit).
(b) Operators who handle instrumentation and control equipment important to safety.

In addition, individuals in positions such as plant manager, head of operations, head of maintenance, directors in technical support and engineering, and certain categories of operator (i.e. other than control room operators, such as fuel operators) should be formally authorized by the operating organization.

7.5. In the assessment of an individual's competence as a basis for authorization, documented and approved criteria should be used. These criteria should include the following:

(a) A knowledge of the established safe working practices and associated regulatory requirements, as appropriate for the assigned duties;
(b) A knowledge of the plant and the safety systems, as relevant to the assigned duties;
(c) Technical, interpersonal, administrative and management knowledge and skills, as appropriate for the assigned duties;
(d) The minimum levels of education, training and experience.

7.6. If an authorized individual moves to a different plant or to a different position in the same plant for which an authorization is also necessary, the individual's authorization should be reassessed before the new position is assumed.

7.7. The need for periodic reauthorization should be considered. An authorization is typically subject to periodic reviews (at intervals of 2–3 years) and can be withdrawn (or not extended), if the competence of the authorized person no longer meets the criteria described in para. 7.5. Renewal or extension of the authorization should, in all cases, be subject to acceptable results of a recent medical examination.

7.8. The need for reauthorization of individuals who are expected to resume authorized duties after an extended period of absence should be assessed. This assessment should take into account changes (e.g. in the plant and in procedures) that have occurred during the period of absence. A graded approach can be applied to such reauthorization, by means of targeted training and assessments.

REFERENCES

[1] INTERNATIONAL ATOMIC ENERGY AGENCY, Safety of Nuclear Power Plants: Commissioning and Operation, IAEA Safety Standards Series No. SSR-2/2 (Rev. 1), IAEA, Vienna (2016).

[2] INTERNATIONAL ATOMIC ENERGY AGENCY, Safety of Nuclear Power Plants: Design, IAEA Safety Standards Series No. SSR-2/1 (Rev. 1), IAEA, Vienna (2016).

[3] INTERNATIONAL ATOMIC ENERGY AGENCY, Leadership and Management for Safety, IAEA Safety Standards Series No. GSR Part 2, IAEA, Vienna (2016).

[4] INTERNATIONAL ATOMIC ENERGY AGENCY, Operational Limits and Conditions and Operating Procedures for Nuclear Power Plants, IAEA Safety Standards Series No. SSG-70, IAEA, Vienna (2022).

[5] INTERNATIONAL ATOMIC ENERGY AGENCY, Modifications to Nuclear Power Plants, IAEA Safety Standards Series No. SSG-71, IAEA, Vienna (2022).

[6] INTERNATIONAL ATOMIC ENERGY AGENCY, The Operating Organization for Nuclear Power Plants, IAEA Safety Standards Series No. SSG-72, IAEA, Vienna (2022).

[7] INTERNATIONAL ATOMIC ENERGY AGENCY, Core Management and Fuel Handling for Nuclear Power Plants, IAEA Safety Standards Series No. SSG-73, IAEA, Vienna (2022).

[8] INTERNATIONAL ATOMIC ENERGY AGENCY, Maintenance, Testing, Surveillance and Inspection in Nuclear Power Plants, IAEA Safety Standards Series No. SSG-74, IAEA, Vienna (in press).

[9] INTERNATIONAL ATOMIC ENERGY AGENCY, Conduct of Operations at Nuclear Power Plants, IAEA Safety Standards Series No. SSG-76, IAEA, Vienna (in press).

[10] INTERNATIONAL ATOMIC ENERGY AGENCY, IAEA Safety Glossary: Terminology Used in Nuclear Safety and Radiation Protection, 2018 Edition, IAEA, Vienna (2019).

[11] INTERNATIONAL ATOMIC ENERGY AGENCY, Milestones in the Development of a National Infrastructure for Nuclear Power, IAEA Nuclear Energy Series No. NG-G-3.1 (Rev. 1), IAEA, Vienna (2015).

[12] INTERNATIONAL ATOMIC ENERGY AGENCY, Establishing the Safety Infrastructure for a Nuclear Power Programme, IAEA Safety Standards Series No. SSG-16 (Rev. 1), IAEA, Vienna (2020).

[13] INTERNATIONAL ATOMIC ENERGY AGENCY, Workforce Planning for New Nuclear Power Programmes, IAEA Nuclear Energy Series No. NG-T-3.10, IAEA, Vienna (2011).

[14] INTERNATIONAL ATOMIC ENERGY AGENCY, IAEA World Survey on Nuclear Power Plant Personnel Training, IAEA-TECDOC-1063, IAEA, Vienna (1999).

[15] INTERNATIONAL ATOMIC ENERGY AGENCY, Systematic Approach to Training for Nuclear Facility Personnel: Processes, Methodology and Practices, IAEA Nuclear Energy Series No. NG-T-2.8, IAEA, Vienna (2021).

[16] INTERNATIONAL ATOMIC ENERGY AGENCY, Experience in the Use of Systematic Approach to Training (SAT) for Nuclear Power Plant Personnel, IAEA-TECDOC-1057, IAEA, Vienna (1998).

[17] INTERNATIONAL ATOMIC ENERGY AGENCY, Managing Human Resources in the Field of Nuclear Energy, IAEA Nuclear Energy Series No. NG-G-2.1, IAEA, Vienna (2009).

[18] FOOD AND AGRICULTURE ORGANIZATION OF THE UNITED NATIONS, INTERNATIONAL ATOMIC ENERGY AGENCY, INTERNATIONAL CIVIL AVIATION ORGANIZATION, INTERNATIONAL LABOUR ORGANIZATION, INTERNATIONAL MARITIME ORGANIZATION, INTERPOL, OECD NUCLEAR ENERGY AGENCY, PAN AMERICAN HEALTH ORGANIZATION, PREPARATORY COMMISSION FOR THE COMPREHENSIVE NUCLEAR-TEST-BAN TREATY ORGANIZATION, UNITED NATIONS ENVIRONMENT PROGRAMME, UNITED NATIONS OFFICE FOR THE COORDINATION OF HUMANITARIAN AFFAIRS, WORLD HEALTH ORGANIZATION, WORLD METEOROLOGICAL ORGANIZATION, Preparedness and Response for a Nuclear or Radiological Emergency, IAEA Safety Standards Series No. GSR Part 7, IAEA, Vienna (2015).

[19] INTERNATIONAL ATOMIC ENERGY AGENCY, Preparation, Conduct and Evaluation of Exercises to Test Preparedness for a Nuclear or Radiological Emergency, EPR-Exercise 2005, IAEA, Vienna (2005).

[20] INTERNATIONAL ATOMIC ENERGY AGENCY, Accident Management Programmes for Nuclear Power Plants, IAEA Safety Standards Series No. SSG-54, IAEA, Vienna (2019).

[21] INTERNATIONAL NUCLEAR SAFETY ADVISORY GROUP, Safety Culture, Safety Series No. 75-INSAG-4, IAEA, Vienna (1991).

[22] EUROPEAN COMMISSION, FOOD AND AGRICULTURE ORGANIZATION OF THE UNITED NATIONS, INTERNATIONAL ATOMIC ENERGY AGENCY, INTERNATIONAL LABOUR ORGANIZATION, OECD NUCLEAR ENERGY AGENCY, PAN AMERICAN HEALTH ORGANIZATION, UNITED NATIONS ENVIRONMENT PROGRAMME, WORLD HEALTH ORGANIZATION, Radiation Protection and Safety of Radiation Sources: International Basic Safety Standards, IAEA Safety Standards Series No. GSR Part 3, IAEA, Vienna (2014).

CONTRIBUTORS TO DRAFTING AND REVIEW

Andersson, O.	Consultant, Sweden
Asfaw, K.	International Atomic Energy Agency
Bassing, G.	Consultant, Germany
Cavellec, R.	International Atomic Energy Agency
Depas, V.	Engie Electrabel, Belgium
Lipar, M.	Consultant, Slovakia
Nikolaki, M.	International Atomic Energy Agency
Noël, M.	European Commission Joint Research Centre, Belgium
Ranguelova, V.	International Atomic Energy Agency
Shaw, P.	International Atomic Energy Agency
Tararin, A.	Rosenergoatom, Russian Federation
Vaišnys, P.	Consultant, Austria

No. 26

ORDERING LOCALLY

IAEA priced publications may be purchased from the sources listed below or from major local booksellers.

Orders for unpriced publications should be made directly to the IAEA. The contact details are given at the end of this list.

NORTH AMERICA

Bernan / Rowman & Littlefield
15250 NBN Way, Blue Ridge Summit, PA 17214, USA
Telephone: +1 800 462 6420 • Fax: +1 800 338 4550
Email: orders@rowman.com • Web site: www.rowman.com/bernan

REST OF WORLD

Please contact your preferred local supplier, or our lead distributor:

Eurospan Group
Gray's Inn House
127 Clerkenwell Road
London EC1R 5DB
United Kingdom

Trade orders and enquiries:
Telephone: +44 (0)176 760 4972 • Fax: +44 (0)176 760 1640
Email: eurospan@turpin-distribution.com

Individual orders:
www.eurospanbookstore.com/iaea

For further information:
Telephone: +44 (0)207 240 0856 • Fax: +44 (0)207 379 0609
Email: info@eurospangroup.com • Web site: www.eurospangroup.com

Orders for both priced and unpriced publications may be addressed directly to:

Marketing and Sales Unit
International Atomic Energy Agency
Vienna International Centre, PO Box 100, 1400 Vienna, Austria
Telephone: +43 1 2600 22529 or 22530 • Fax: +43 1 26007 22529
Email: sales.publications@iaea.org • Web site: www.iaea.org/publications